Student Workbook Volume II
to Accompany

Freedman, Ruskell, Kesten, and Tauck's
College Physics

First Edition

Mark Kanner
CUNY Graduate Center, CUNY City College

Jason Bryslawskyj
CUNY Graduate Center, CUNY Baruch

W.H. Freeman and Company
New York

A Macmillan Higher Education Company

Student Workbook Volume I: 1-4641-4957-7
Student Workbook Volume II: 1-4641-4958-5

© 2014 by W. H. Freeman and Company

All rights reserved.

Printed in the United States of America

First printing

W. H. Freeman and Company
41 Madison Avenue
New York, NY 10010
Houndmills, Basingstoke RG21 6XS England

www.whfreeman.com

CONTENTS

    Preface   *iv*

Volume II

16   Electrostatics I: Electric Charge, Forces, and Fields, **1**

17   Electrostatics II: Electric Potential Energy and Electric Potential, **13**

18   Electric Charges in Motion, **29**

19   Magnetism, **43**

20   Electromagnetic Induction, **59**

21   Alternating-Current Circuits, **69**

22   Electromagnetic Waves, **83**

23   Wave Properties of Light, **90**

24   Geometrical Optics, **107**

25   Relativity, **123**

26   Quantum Physics and Atomic Structure, **136**

27   Nuclear Physics, **148**

28   Particle Physics and Beyond, **160**

PREFACE

Taking your first physics course can be an intimidating experience. Perhaps you have heard from family and friends that physics is hard, confusing, or even *boring*! While it is impossible to verify these subjective descriptions, we *can* say that physics is an endeavor that takes time and energy. If you were interested in playing the violin and someone handed you one, you wouldn't expect to be able to play it right away; learning physics is no different. It takes many hours of patience and effort to see results, but fortunately if you put the time in, you do get better. This workbook is aimed at making this process progress as quickly as possible.

The *Student Workbook to Accompany College Physics* is meant as a supplement to *College Physics* and can be used in many different ways. Since each chapter and section of the workbook closely mirror *College Physics*, this *Student Workbook* can help develop many of the skills required for success in learning introductory physics. One use for the *Student Workbook* is as an additional pedagogical resource for professors teaching a course using the book. All the problems match the curriculum in *College Physics,* so they can be used as exam/quiz material that targets specific sections in the textbook.

The workbook is also useful for students when they encounter difficulty in solving problems in a section of *College Physics*. The problems in the *Student Workbook* are structured so that the student is carefully led through the steps in a question. If you are overwhelmed or confused by a topic, this scaffolded approach can help by walking you through all of the steps. After you gain some insight from solving problems that have been parsed out for you, it will be easier to solve full homework problems.

We sincerely hope you find the problems contained in this book useful to your understanding of physics.

Sincerely,

Mark Kanner & Jason Bryslawskyj

Chapter 16 Electrostatics I: Electric Charge, Forces, and Fields

Section 16-1 Electric forces and electric charges are all around you—and within you

Goal: Explain why studying electric phenomena is important.

Concept Check
1. What is electrostatics?

Section 16-2 Matter contains positive and negative electric charge

Goal: Describe how objects acquire a net electric charge.

Concept Check
1. What does it mean to say something is quantized?

2. A hydrogen atom has on proton and one electron. What is the charge measured from a hydrogen atom from far away?

Problem-Solving Review
In an experiment, Erin holds two objects. Object A has a charge of -4.81×10^{-7} C and object B has a charge of 6.41×10^{-7} C. She touches the objects together.

Set Up
1. Will the charge q_a in object A increase, decrease, or stay the same after the objects are touched together?

2. Will the charge q_b in object B increase, decrease, or stay the same after the objects are touched together?

Solve
1. Roughly how many more electrons are there than protons in the object A before the two objects come in contact?

2. Roughly how many more protons are there than electrons in the object B before the two objects come in contact?

3. After Erin touches the objects together, about 5.00×10^{11} electrons transfer from object A to object B. What is the net charge in object B?

Reflect

1. What happens to the force between object A and object B after they touch?

Chapter 16

Section 16-3 Charge can flow freely in a conductor but not in an insulator

Goal: Recognize the differences between insulators, conductors, and semiconductors.

Concept Check

1. Wires often are copper surrounded by plastic. Is plastic an insulator or a conductor?

2. If you were to measure the velocity of an electron in a copper wire and the velocity of an electron in the plastic sheath around it, which would be greater?

3. What are the charge carriers in metals? What are the charge carriers in biological systems?

Section 16-4 Coulomb's law describes the force between charged objects

Goal: Use Coulomb's law to quantitatively describe the force that one charged particle exerts on another.

Concept Check

1. Which quantity is larger, the gravitational constant G or the Coulomb constant k?

Problem-Solving Review

Particle A with a charge of 5.00×10^{-4} C is held suspended from the ceiling 20.0 m from the ground. A second particle, particle B, with a charge of -7.00×10^{-4} C and a mass of 1.30 kg needs to be placed in a position so that it is in equilibrium with the first particle and gravity.

Set Up

1. What are the forces acting on particle B? Write Newton's first law for the particle in terms of the forces acting on particle B.

2. What is the magnitude of the gravitational force acting on particle B?

Solve

1. How far above the floor should you place particle B so that it is in equilibrium?

2. After placing particle B in equilibrium with particle A, particle A's charge is increased to -5.00×10^{-4} C. What is the initial net force on the particle B? What is its initial acceleration?

Reflect

1. If you halved the distance between the two charged particles how much would the force between them change? Assume that particle A is positively charged and particle B is negatively charged.

Section 16-5 The concept of electric field helps us visualize how charges exert forces at a distance

Goal: Explain the relationship between electric force and electric field.

Concept Check

1. Sketch the electric field lines for a positively charged particle. Do the field lines point toward or away from the particle?

2. Sketch the electric field lines for a negatively charged particle. Do the field lines point toward or away from the particle?

Problem-Solving Review

Consider two negatively charged particles on the x-axis, $q_1 = -2.00 \times 10^{-4}$ C at $x = 1.00$ cm and $q_2 = -2.00 \times 10^{-4}$ C at $x = -1.00$ cm.

Set Up

1. Sketch the charge configuration. Draw and label a point p_1 at $(0,0)$ and a point p_2 at $(0,5)$. We will observe the electric field of the charge configuration at these points.

Solve
1. What is the magnitude and direction of the electric field at p_1
2. What is the magnitude and direction of the electric field at p_2 from charge q_1?

3. What is the magnitude and direction of the electric field at p_2 from charge q_2?

4. What is the total electric field at p_2?

Reflect
1. If you placed a particle with charge 1.50 C at point p_2, what would the force on the particle be?

Section 16-6 Gauss's law gives us more insight into the electric field

Goal: Describe the connection between enclosed charge and electric flux described by Gauss's law.

Concept Check
1. Explain why the electric flux for a closed surface is always zero if there is no net charge inside.

Problem-Solving Review
Lauren wishes to measure the charge of a particle. She finds that the electric flux through a closed loop of radius 0.0500 m that is 1.15 m from a point charge is measured to be 3.62×10^6 Nm²/C. Assume Lauren holds the loop perpendicular to the field lines from the point charge at a constant distance of 1.15 m from the mystery charge.

Set Up
1. What is the area enclosed by the loop?

2. What is the surface area of the sphere around the point charge at 1.15 m?

3. What proportion of the total surface area is enclosed by the loop?

Solve
1. What is the magnitude of electric field through the loop?

Chapter 16

2. Use your answer in question 3 of the setup to find the total electric flux through an enclosed surface 1.15 m from the charged particle.

3. Use the total electric flux and Gauss's law to find the enclosed charge.

Reflect

1. If Lauren took her measurement at 2.30 m, would this change the total electric flux?

Section 16-7 In certain situations, Gauss's law helps us to calculate the electric field and to determine how charge is distributed

Goal: Apply Gauss's law to symmetric situations and to the distribution of excess charge on conductors.

Concept Check

1. For what types of charge distributions can Gauss's law be used to calculate electric fields?

2. For a large charged sheet, how does the electric field change with the distance from the sheet?

Problem-Solving Review

A thin hollow conducting sphere with radius $R = 0.012$ m has a surface charge density of 0.0200 C/m².

Set Up

1. What is the surface area of the sphere?

2. What is the total charge on the sphere?

Solve

1. What is q_{enc} for a Gaussian sphere with $r = 0.0100$ m? Assume the center of the charged spherical shell is the center of the Gaussian surface. What is the total electric flux through the Gaussian sphere? What is the electric field inside the shell?

2. What is q_{enc} for a Gaussian sphere with $r = 1.00$ m? Assume the center of the charged spherical shell is the center of the Gaussian surface. What is the total electric flux through the Gaussian sphere?

3. What is the electric field at a distance of 1.00 m from the center of the spherical shell?

4. What is q_{enc} for a Gaussian sphere with $r = 5.00$ m? Assume the center of the charged spherical shell is the center of the Gaussian surface. What is the total electric flux through the Gaussian sphere?

5. What is the electric field at a distance of 5.00 m from the center of the spherical shell?

Reflect

1. If the charge were spread throughout the sphere equally, would the electric field be greater than, less than, or equal to the electric field measured in question 1 of Solve?

2. If the charge were spread throughout the sphere equally, would the electric field be greater than, less than, or equal to the electric field measured in question 3 of Solve?

Chapter 17 Electrostatics II: Electric Potential Energy and Electric Potential

Section 17-1 Electrical energy is important in nature technology and biological systems

Goal: Explain the significance of energy in electrostatics

Concept Check

1. What is electric potential energy?

2. What is electric potential?

Section 17-2 Electric potential energy changes when a charge moves in an electric field

Goal: Discuss how the work done on a charged particle by the electric field relates to changes in electric potential energy.

Concept Check
1. How do you change the potential energy of a charged object in an electric field?

Problem-Solving Review
Consider an electron at point (0,0) with an initial velocity of 1.34×10^6 m/s traveling in the $-x$ direction along the x-axis in an electric field pointed in the positive x direction with magnitude 4.43 N/C.

Set Up
1. Sketch the particle and the field lines of the electric field. Label the velocity of the particle.

2. What is the initial kinetic energy of the particle?

Solve

1. Write the conservation of energy equation for the particle in the electric field from the initial position until the particle's final velocity is 0.00 m/s.

2. What is the position of the particle when its velocity reaches 0.00 m/s?

3. Assume that when the particle reaches the position found in question 2, the electric field is switched off (so there is no external electric field) and another charged particle with charge +5.47 nC is brought from very far away to the point (0,−2). Find the electric potential energy of this system.

Reflect

1. How much work was done on the second charge to move it into position?

Chapter 17

Section 17-3 Electric potential equals electric potential energy per charge

Goal: Explain the difference between electric potential and electric potential energy.

Concept Check

1. What are the units of electric potential energy and electric potential?

2. What is the difference between electric potential energy and electric potential?

Problem-Solving Review

To create x-rays (like those used in a CT scan at the hospital), electrons are pulled off a metal plate in a tube by a large electrical potential and accelerated until they crash into a metal target. The energy from the impact is then converted into electromagnetic radiation. We will calculate the change in electric potential of the electrons inside the tube, considering they arrive at the target with a velocity of 2.24×10^6 m/s.

Set Up

1. What is the kinetic energy of the electrons immediately before impact with the target?

2. What is the change in electrical potential energy of an electron in the tube between its initial start from rest and just before it strikes the target?

Solve

1. What is the electric potential between the plate and the target?

2. What is the electric potential halfway between the plate and the target?

Reflect

1. If the charges were positrons instead of electrons, how would the electric potential have to be changed so the positrons would arrive at the target with the same velocity?

Chapter 17

Section 17-4 The electric potential has the same value everywhere on an equipotential surface

Goal: Recognize why equipotential surfaces are perpendicular to electric field lines

Concept Check
1. What is an equipotential surface?

2. What is the significance of field lines being perpendicular to an equipotential surface?

Problem-Solving Review
Consider a thin conducting spherical shell with radius $R = 1.50$ m holding charge $q = 1.00 \times 10^{-6}$ C.

Set Up
1. Sketch the electric field lines for the conductor both inside and outside the shell.

2. What is the electric field inside the shell?

Solve

1. Sketch the electric field lines for the conductor both inside and outside the shell.

2. In the drawing above, draw four equipotential curves around the conductor.

3. Find the potential at $R_1 = 1.51$ m, and $R_2 = 5.00$ m.

Reflect

1. Should the equipotential curves be spaced closer together nearer to the shell or further away?

2. Inside the shell, does the electric potential increase, decrease, or remain constant as you move from the center to the edge?

Chapter 17

Section 17-5 A capacitor shares equal amounts positive and negative charge

Goal: Explain what is meant by capacitance, and describe how the capacitance of a parallel-plate capacitor depends on the size of the plates and their separation.

Concept Check
1. What is capacitance?

2. What two factors affect a capacitors capacitance?

Problem-Solving Review
Ken wants to design a circular capacitor consisting of two circular plates with a radius of 0.00250 m that can hold 8.65×10^{-9} C. The capacitor will have an electric potential of 5.00 V across its plates. Ken needs to determine how far apart to put the plates to get it to operate to specifications.

Set Up
1. What is the area of the plate in the capacitor?

Solve

1. What capacitance is required to store the charge $q = 8.65 \times 10^{-9}$ C at 5.00 V?

2. For the capacitor to work correctly, how far away should the plates be?

3. If you doubled the electric potential in the circuit, how much more charge would the capacitor store?

Reflect

1. What is the charge on the negative plate of the capacitor when it's fully charged?

Chapter 17

Section 17-6 A capacitor is a store house of electric potential energy

Goal: Calculate the electric energy stored in a capacitor.

Concept Check
1. Assuming a fixed capacitance, what two quantities affect the electric potential energy stored on a capacitor?

Problem-Solving Review
For fast accelerations, energy stored in a capacitor will be used to briefly power an electric car with a mass of 1000.00 kg. The electric car needs to be able to reach a velocity of 50.0 m/s starting from rest. We will calculate the capacitance of the capacitor required for this acceleration, considering that the capacitor is charged by applying 7.00 kV.

Set Up
1. What is the kinetic energy of the car at 50.0 m/s?

2. Write down the conservation of energy equation from when the car is at rest and all the energy is stored in the capacitor, and when the car is traveling at 50.0 m/s.

Solve

1. What is the electrical energy stored in the capacitor before it is discharged, assuming that all the electrical energy is converted to kinetic energy?

2. What is the amount of charge stored in the capacitor before it is discharged?

3. What is the capacitance of the capacitor?

Reflect

1. If we increased the electric potential used to charge the capacitor, would the potential electrical energy stored on the capacitor increase, decrease, or stay the same?

Chapter 17

Section 17-7 Capacitors can be combined in series or in parallel

Goal: Explain how to treat capacitors attached in series and in parallel as a single equivalent capacitance.

Concept Check

1. Sketch three capacitors with capacitance C_1, C_2, and C_3, respectively, in series.

2. Sketch three capacitors with capacitance C_1, C_2, and C_3, respectively, in parallel.

Problem-Solving Review

Consider the following circuit diagram with C_1 = 5.00 mF, C_2 = 7.00 mF, and C_3 = 13.0 mF. The battery is rated at 5.00 V.

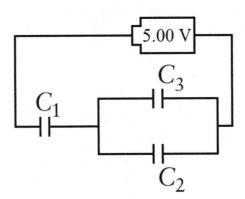

Set Up
1. What is the equivalent capacitance of capacitors C_2 and C_3?

Solve
1. Use your result from the Set Up to find the equivalent capacitance of the entire circuit.

2. What is the total charge q held in all the capacitors of the circuit, assuming they are all fully charged? (*Hint:* This can be found from the equivalent capacitance of the circuit.)

3. What is the voltage across capacitor 1?

4. Find q_1, the amount of charge in capacitor 1.

5. Find the charge stored in capacitor 2 and capacitor 3.

Reflect

1. What is the potential across C_2?

2. What is the potential across C_3?

Section 17-8 Placing a dielectric between the plates of a capacitor increases the capacitance

Goal: Describe how the capacitance of a capacitor increases when an insulating material other than a vacuum is placed between the plates of the capacitor.

Concept Check
1. What is a dielectric?

2. Describe what happens to the molecules inside a dielectric when it is placed between the plates of a charged capacitor.

Problem-Solving Review
Two capacitors in series with capacitance $C_1 = 7.33$ mF and $C_2 = 2.89$ mF are connected to a battery that has a potential of 12.0 V. After the capacitors are charged, a dielectric with $\kappa = 3.15$ is slid inside capacitor 2.

Set Up
1. Before the dielectric is placed between the plates, what is the equivalent capacitance of the circuit?

2. Will the dielectric increase, decrease, or not change the effective capacitance of the circuit?

Solve

1. What is C_2 after the dielectric is inserted between the two plates?

2. What is the equivalent capacitance of the system?

3. What is the ratio of the electric field in capacitor 2 before the dielectric was inserted to the electric field between the two plates of capacitor 2 after it was inserted?

Reflect

1. If a dielectric material was placed between the plates of the first capacitor, would the equivalent capacitance of the system increase, decrease, or stay the same?

Chapter 18 Electric Charges in Motion

Section 18-1 Life on Earth and our technological society are only possible because of charges in motion

Goal: Recognize why moving electric charges are important.

Concept Check

1. Give a few real world examples of process that involve moving charges.

2. Explain the difference between current, resistance, and source of emf.

Chapter 18

Section 18-2 Electric current equals the rate at which charge flows

Goal: Explain the meaning of current and drift speed, and the difference between direct and alternating current.

Concept Check
1. Explain the relationship between electric fields and current.

2. Explain the difference between current and drift speed.

3. Explain the difference between alternating and direct current.

Problem-Solving Review
A current of 0.750 A is driven into a rectangular parallel plate capacitor. The capacitor has dimensions of 1.00 cm × 5.00 cm and a distance of 0.0500 mm between plates. Let's calculate how long it takes for the capacitor to reach 25.0 V.

Set Up
1. Find an expression for the voltage between the plates of the capacitor in terms of the given dimensions, the accumulated charge, and ε_0.

2. Find an expression for the time it takes to change the amount of charge on the capacitor in terms of the amount of accumulated and the applied current.

Solve

1. Using the equation you found in question 1 of the Set Up, calculate the amount of charge that needs to accumulate in order for there to be a voltage of 25.0 V between the plates of the capacitor.

2. Using the equation you found in question 2 of the Set Up, calculate the time it takes the capacitor to reach 25 V.

Reflect

1. The wires carrying current in this circuit have a radius of 0.500. What is the drift speed of the electrons? (Assume there are 8.49×10^{28} free electrons per meter in the copper wires and the charge per electron is 1.60×10^{-19} C.)

Section 18-3 The resistance to current through an object depends on the object's resistivity and dimensions

Goal: Describe the relationships among voltage, current, resistance, and resistivity for charges moving in a wire.

Concept Check

1. Explain how resistance arises in a current carrying wire.

2. If the thickness of a wire is doubled, will its resistance increase, decrease, or stay the same? What would happen if its length were doubled?

Problem-Solving Review

Karl runs a 10.0-ft ethernet cable from his computer to his network router. There are eight wires in an ethernet cable. Each wire in the cable is made of copper and has a radius of about 0.255 mm. On one of the wires, a voltage of 0.250 V is applied. Let's calculate the resistance in one of the wires and the current flowing in it. Assume that the resistivity of copper is 1.725×10^{-8} Ωm.

Set Up

1. What is the length of the ethernet cable in m?

2. Write an expression for the resistance of the wire in terms of its length, cross sectional area, and resistivity.

3. Write an expression for the current flowing in the wire in terms of the applied voltage and the resistance of the wire.

Solve
1. Calculate the cross-sectional area of the wire.

2. Calculate the resistance of the wire.

2. Calculate the amount of current flowing in the wire.

Reflect
1. What would the resistance of the wire be if its radius were halved?

2. How much current would be flowing in this case?

Section 18-4 Resistance is important in both technology and physiology

Goal: Calculate the resistance of a resistor and the current that a given voltage produces in that resistor.

Concept Check
1. Explain the importance of electrical resistance for cellular membranes.

2. Explain how a nerve signal propagates along an axon.

Problem-Solving Review
A cellular membrane contains a channel that allows the flow of K⁺ ions across the membrane. The channel has a radius of 0.500 nm and the solution inside the channel has a resistivity of 0.520 Ωm. Let's explore the electrical properties of the channel.

Set Up
1. What is 1.00 nm in units of meters?

2. Calculate the cross-sectional area of the channel.

3. Write an expression for the resistance of the channel in terms of its resistivity, length, and area.

Solve
1. How long is the channel if its resistance is 1.00×10^{10} Ω?

2. If there is a current of 0.751 pA flowing through the channel, what is the voltage across the channel?

3. How many ions are passing through the channel per second?

Reflect

1. What is the magnitude of the electric field causing the potential difference across the channel?

Section 18-5 Kirchhoff's rules help us to analyze simple electric circuits

Goal: Discuss Kirchhoff's rules and how to apply them to single-loop and multiloop circuits.

Concept Check
1. Explain why the sum of changes in potential around a closed loop is zero.

2. Explain why the emf provided by a battery in a circuit is not equal to the voltage that is printed on the battery's label.

3. Explain why the sum of the currents flowing into a junction must equal the sum of the currents flowing out of the junction.

Problem-Solving Review
In the circuit pictured here, a large battery is connected in series to a single resistor and a parallel network of three resistors. The battery provides an emf ε of 25.0 V and has an internal resistance r of 0.100 Ω. The values of the four resistors are $R_1 = 1.00$ kΩ, $R_2 = 220$ Ω, $R_3 = 470$ Ω, and $R_4 = 27.0$ kΩ. Let's calculate the voltage across R_3 and the current through R_4.

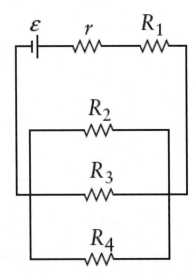

Set Up

1. First, we will reduce the parallel network to a single resistor with equivalent resistance labeled $R_\parallel$. The equivalent resistance $R_\parallel$ is the value of a single resistor that has the same resistance as R_1, R_2, and R_3 in parallel. Write an expression for $\frac{1}{R_\parallel}$ in terms of R_1, R_2, and R_3.

2. Draw a new schematic of the circuit, replacing the parallel network of R_1, R_2, and R_3 with a single resistor labeled $R_\parallel$.

3. What is the relationship between $V_\parallel$ (the voltage across the entire parallel network), V_2 (the voltage across R_2), V_3 (the voltage across R_3), and V_4 (the voltage across R_4)?

4. Now we will find the value of R_{eq}, a single resistor that has the same resistance as r, R_1, and $R_\parallel$ in series. Write an expression for R_{eq}, in terms of r, R_1, and $R_\parallel$.

5. Now draw another schematic, this time replace r, R_1, and $R_{||}$ with a single resistor labeled R_{eq}.

6. What is the relationship between i_{eq} (the current flowing through R_{eq}), i_r (the current flowing through r), i_1 (the current flowing through R_1), and $i_{||}$ (the current flowing through $R_{||}$)?

Solve
1. What is the value of R_{eq}?

2. What is the value of i_{eq}, the current flowing through R_{eq}?

3. What is the value of $V_{||}$, the voltage across $R_{||}$?

4. What is the value of V_3, the voltage across R_3?

5. What is the value of i_4, the current flowing through R_4?

Reflect
1. What is the value of $i_2 + i_3$?

Section 18-6 The rate at which energy is produced or taken in by a circuit element depends on current and voltage

Goal: Calculate the power into or out of a circuit element.

Concept Check
1. Explain the difference between voltage and electrical potential energy.

2. Explain why there is a decrease in electrical potential energy when charges move across a resistor.

Problem-Solving Review
A 30.0 V battery is connected to four resistors as shown in the schematic here. The values of the four resistors are: $R_1 = 330\ \Omega$, $R_2\ 470\ \Omega$, $R_3 = 1.00\ k\Omega$, and $R_4 = 2.20\ k\Omega$. Let's calculate power is dissipated across R_1. You may ignore the internal resistance of the battery.

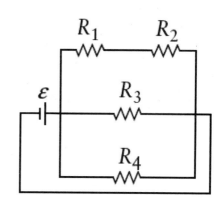

Set Up
1. Let's replace R_1 and R_2 with a single resistor of equivalent resistance called R_{eq}. Since R_1 and R_2 are in series, what is the value of R_{eq}?

2. Redraw the circuit schematic replacing R_1 and R_2 with a single resistor of equivalent resistance called R_{eq}.

3. Now we will replace the parallel network of R_{eq}, R_3, and R_4 with a single resistor of resistance $R_{||}$. What is the value of $R_{||}$?

4. Redraw the circuit schematic replacing R_{eq}, R_3, and R_4 with a single resistor labeled $R_{||}$.

Solve

1. What is the voltage across $R_{||}$?

2. What is the current flowing through R_{eq}? What is the current flowing through R_1?

3. How much power is dissipated across R_1?

Reflect

1. How much power is consumed by the entire circuit?

Section 18-7 A circuit containing a resistor and capacitor has a current that varies with time

Goal: Explain what happens when a capacitor in series with a resistor is charged or discharged.

Concept Check

1. Describe what happens to the current flowing through a capacitor while it is being charged.

2. Describe how a cellular membrane can act as a capacitor and how it can be represented by an RC circuit.

Problem-Solving Review

Charlie has two identical capacitors, but he doesn't know their values. He connects them to a battery, a 1.00 kΩ resistor and a switch as shown in the circuit schematic below. He connects the capacitors in parallel and uses the battery to charge them. After charging the capacitors, Charlie flips the switch and measures how long it takes them to discharge in order to find their capacitance. We will consider the circuit when the switch is flipped to the discharged position. Immediately after he flips the switch, the capacitors hold a combined charge of 9.00 × 10^{-4} C. Charlie finds that the charge on the capacitor falls to half its value in 12.5 ms.

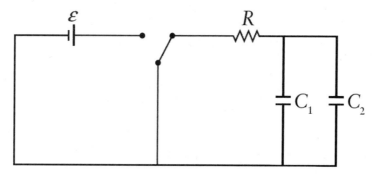

Set Up
1. The capacitors C_1 and C_2 both have the same capacitance and are combined in parallel. What is the value of a single capacitor C_{eq} that has an equivalent capacitance to the parallel capacitors in terms of the value of C_1 and C_2?

2. Redraw the schematic replacing the two capacitors C_1 and C_2 with a single capacitor C_{eq} that has an equivalent capacitance.

3. Write an expression for the charge of the capacitor as a function of time $q(t)$ in terms of q_{max}, t, R, and C_{eq}.

Solve
1. What is $q(t = 12.5\text{ms})$?

2. Use the expression you found in question 3 of the Set Up to find the value of RC_{eq}.

3. What is the capacitance of each capacitor?

Reflect
1. How much current is flowing through the circuit at $t = 25.0$ ms?

Chapter 19 Magnetism

Section 19-1 Magnetic forces are interactions between two magnets

Goal: Recognize that magnetic forces can act over large distances.

Concept Check

1. What is one similarity between magnetic forces and electric forces?

2. Name three devices that use magnets to operate.

Section 19-2 Magnetism is an interaction between moving charges

Goal: Recognize that magnetism is fundamentally an interaction between moving electric charges.

Concept Check

1. What is magnetism?

2. What two conditions are required for an object to experience a magnetic force?

3. Moving charge creates a magnetic field. In a magnet, where does the moving charge occur?

Section 19-3 A moving point charge can experience a magnetic force

Goal: Calculate the magnitude and direction of the magnetic force on a charged particle.

Concept Check
1. What four quantities affect the magnitude of the force exerted on a moving charged particle?

2. What are the SI units of a magnetic field?

Problem-Solving Review
Consider the case of an electron travelling in the $+y$ direction with a velocity of 4.5×10^4 m/s that travels through a magnetic field of strength 2.11×10^{-3} T pointing in the $+z$ direction (up out of the plane of the page).

Set Up
1. Make a sketch showing the vectors for v, B, and F.

Solve
1. What is the magnitude of the magnetic force on the electron?

46   Chapter 19

2. What is the direction of the magnetic force on the electron?

3. The magnetic force acting on the particle will cause the particle to exhibit a circular trajectory. What is the centripetal acceleration of the electron?

4. What is the radius of the circle in which the electron will travel?

Reflect

1. If the strength of the magnetic field were doubled, how would this change the magnetic force acting on the particle?

Section 19-4 A mass spectrometer uses magnetic forces to differentiate atoms of different masses

Goal: Describe how a mass spectrometer uses magnetic fields to sort atoms according to their mass.

Concept Check
1. Describe how a mass spectrometer works.

Problem-Solving Review

Strontium-90 is a radioactive isotope that emits electrons as it decays into other elements. It is possible to measure the velocity of these electrons using a device similar to a mass spectrometer. When the electrons leave the sample, they enter a magnetic field and a force is exerted on them. Subsequently, their direction is changed until they run into the detector. The detector can slide along the x-axis, and its distance from the strontium source is indicated by d.

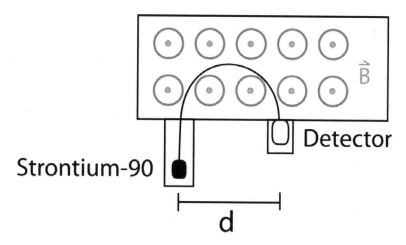

Chapter 19

Assume the strength of the magnetic field is 5.20×10^{-4} T. The mass of an electron is 9.11×10^{-31} kg and the charge is 1.60×10^{-19} C.

Set Up

1. Which way must the magnetic field be pointing for the particle to follow the given trajectory?

2. To use this apparatus as a velocity selector, an electric field could be applied to the particle once it had entered the magnetic field. In what direction would the electric field be?

Solve

1. If the detector is placed at $d = 0.025$ m, what is the velocity of the measured electron? Assume the electric field is off.

2. Assuming the electric field was turned on, what would be the magnitude of the electric field have to be for the electron to go straight?

Reflect

1. How would changing the mass of a particle in a velocity selector affect the particle's velocity?

Section 19-5 Magnetic fields exert forces on current carrying wires

Goal: Calculate the magnetic force on a current-carrying wire.

Concept Check
1. What four quantities determine the strength of the force on a current-carrying wire?

2. Name two devices that use the force generated on a current-carrying wire by a magnetic field in their operation.

Problem-Solving Review
MRI machines use very strong magnets to acquire images. Due to the strength of the magnetic field, very strong forces can be exerted on the current-carrying wires inside the machine. We will consider a thin wire that can withstand forces of up to 80.0 N on any section less than 0.100 m, and has to carry a current of 132 A in a 7.00 T magnetic field to determine if any forces are exerted on the wire that could cause it to fail. The wire can be placed in a variety of angles with respect to the magnetic field.

Set Up
1. What angle between the magnetic field and the current in the wire will give the minimum force on the wire?

2. What angle between the magnetic field and the wire will give the maximum force on the wire?

Solve

1. What is the minimum force on the wire from the magnetic field?

2. What is the maximum force on the wire from the magnetic field?

3. What is the maximum angle at which the magnetic field can be before the wire breaks?

Reflect

1. What is the drift velocity of an electron in the wire if the radius of the wire is 0.0150 m and there are $n = 8.58 \times 10^{28}$ charges per unit volume?

Section 19-6 A magnetic field can exert a torque on a current loop

Goal: Explain why a current loop in a uniform magnetic field experiences a net torque but zero net force.

Concept Check
1. Explain in words why the net force on a current carrying coil from a uniform magnetic field is zero.

Problem-Solving Review
Consider a rectangular loop with sides of length 8.15 cm and 5.67 cm in a uniform magnetic field of 3.00 T in the +x direction rotating about the z-axis. The loop has 1.89 A of current running through it in a clockwise direction.

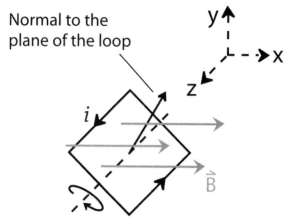

Set Up
1. Write an equation for the torque on the loop.

2. What is the net force on the rectangular loop?

3. Will the loop rotate clockwise or counterclockwise about the z-axis?

Solve
1. What is the torque on the loop when it is at a 15.0-degree angle to the magnetic field?

2. What is the torque on the loop when it is at a 50.0-degree angle to the magnetic field?

3. What is the torque on the loop when it is at an 83.0-degree angle to the magnetic field?

Reflect

1. What effect would reversing the direction of the current have on the rotation of the wire?

Section 19-7 Ampère's law describes the magnetic field created by current-carrying wires

Goal: Describe the principle of Ampère's law and how to use it.

Concept Check
1. What is Ampère's law?

2. Explain why Ampère's law isn't used to measure the magnetic field near current loops.

Problem-Solving Review
An electron moves in the +x direction at 5×10^4 m/s. It is 1.12 m on the x-y plane from a wire running along the x-axis that is carrying 3.00 A of current in the –x direction. We will find the force acting on the electron from the current in the wire.

Set Up
1. Draw the wire and the magnetic field lines. Be sure to indicate the direction of the magnetic field.

2. What is the magnetic field at the location of the electron?

Solve

1. What is the direction of the force exerted on the electron by the magnetic field?

2. What is the magnitude of the force exerted on the electron by the magnetic field?

Reflect

1. What would be the force on the electron if the wire were replaced by a solenoid of radius 0.010 m?

Section 19-8 Two current-carrying wires exert magnetic forces on each other

Goal: Calculate the magnetic force that parallel current-carrying wires exert on each other.

Concept Check

1. Do two wires carrying current in the same direction attract or repel each other?

2. Do two wires carrying current in the opposite direction attract or repel each other?

Problem-Solving Review

A rail gun is a device that uses large magnetic fields to power conducting projectiles. The rail gun has many useful applications in space travel and arms. It has even been suggested that you could place an astronaut inside the projectile of a large rail gun and launch them into space! Consider the rail gun diagram below.

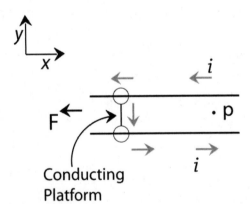

Current runs along the top wire, through a moveable conducting platform, and then traverses the bottom wire as shown above. We will calculate the forces that the two current-carrying wires exert on each other as the rail gun fires. In this case, we will neglect the effects of the conducting platform. The two wires are a distance of 1.11×10^{-1} m apart.

Set Up
1. What is the direction of the magnetic field from both wires at point p in the *x-y* plane?

2. Rail guns require large currents. If $I = 3.00 \times 10^5$ A, what would be the magnitude of the magnetic field at point p? (*Hint:* Remember to include the magnetic field from both wires!)

Solve
1. What is the magnitude and direction of the magnetic field from the top wire at any point on the bottom wire?

2. What is the magnitude and direction of the magnetic field from the bottom wire at any point on the top wire?

3. What is the force per unit length of the top wire on the bottom wire?

4. What is the force per unit length of the bottom wire on the top wire?

Reflect

1. Explain how the force per unit length of the top wire on the bottom wire and the force per unit length of the bottom wire on the top wire satisfy Newton's third law.

Chapter 20 Electromagnetic Induction

Section 20-1 The world runs on electromagnetic induction

Goal: Explain the importance of electromagnetic induction.

Concept Check

1. Explain the conditions under which an electric field can arise from a magnetic field.

2. Using the concept of electromagnetic induction, explain how a credit-card reader works.

Section 20-2 A changing magnetic flux creates an electric field

Goal: Describe what is meant by a motional emf and an induced emf.

Concept Check
1. Explain Faraday's law of induction.

2. A loop of wire is placed in a uniform magnetic field. The magnetic field is pointing in the positive z direction. If the loop is given a velocity in the negative in the z direction, will there be an induced emf? Explain.

Problem-Solving Review
A loop of copper wire is bent into a partial circle, with two ends forming a straight section as show in the diagram below. The ends of the straight section are parallel and are 4.00 cm apart. Along them, a copper rail is free to move back and forth, forming a closed loop. The entire loop is set at an angle of 30.0° with the horizontal. A 2.00 Tesla uniform magnetic field is applied in the vertical direction. Starting from the outermost end, the copper rail is pushed along the straight section with a velocity v, causing an emf of 10.0 V to arise. Let's use Faraday's law of induction to calculate the velocity of the copper rail. (You may ignore the resistance of the loop.)

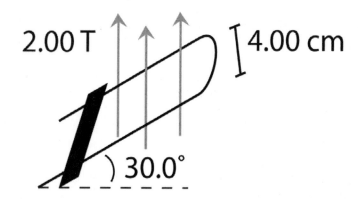

Set Up

1. Write an expression for the amount of magnetic flux passing through the copper wire loop in terms of the cross-sectional area of the loop A, the strength of the magnetic field B, and the angle θ between a perpendicular to the loop and the magnetic field.

2. How large is the angle θ?

3. Do you know the cross-sectional area of the loop? Is this needed to solve the problem?

4. Write an expression for the emf induced by the change in magnetic flux in terms of the change in magnetic flux and the change in time.

5. Does the magnitude of the magnetic field B or the angle θ depend on the time at which it is observed? How about the cross-sectional area A?

Solve

1. Write an expression for the change in cross-sectional area in terms of the length of the copper rail, the velocity at which it is being pushed, and the change in time.

2. Use this expression to find and expression for the induced emf in terms of the velocity of the copper rail, the strength of the magnetic field, and the length of the rail.

3. What is the velocity of the copper rail?

Reflect

1. Would increasing the speed of the copper rail, increase, or decrease the amount of induced emf?

Section 20-3 Lenz's law describes the direction of the induced emf

Goal: Explain what determines the magnitude and direction of an emf in a circuit with a changing magnetic flux.

Concept Check

1. If you hold a closed loop of wire horizontally and drop a bar magnet through it, an emf will be induced in the loop. If the magnetic field of the bar magnet is pointing up, in which direction will the induced magnetic field be pointing when the bar magnet has just passed through the loop?

2. Explain how eddy currents can be used to electrically slow down the wheels of a car.

Problem-Solving Review

A rectangular loop of copper wire is inside a uniform 4.00 T magnetic field as shown in the diagram below. The magnetic field is pointing in the positive z direction (out of the book). The wire is 9.00 cm by 3.00 cm and is connected to a small lamp that has a resistance of 3.00 Ω. Let's calculate how much power will be delivered to the lamp if the loop is pulled out of the magnetic field at 10.0 cm/s. The external magnetic field has straight edges, and the loop is pulled out of one of these straight edges.

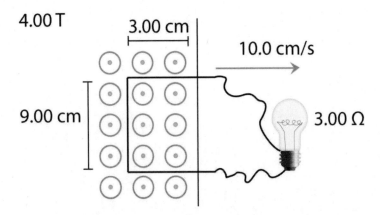

Set Up

1. As the loop is pulled out of the magnetic field, is the magnetic flux through the loop increasing or decreasing?

2. In which direction is the induced magnetic field pointing?

3. What is the direction of the induced current in the loop?

4. Write an expression for the emf induced in the loop in terms of the change in magnetic flux through the loop and the change in time.

5. Write an expression for the power delivered to the lamp in terms of the emf induced in the loop and the resistance of the lamp.

6. As the loop is pulled out of the magnetic field, the area inside the loop that is overlapping the magnetic field decreases. How does the length of this overlapping area change with time as the loop is pulled? How does the width change in time?

Solve

1. Write an expression for the change in the area of the loop that is inside the magnetic field. This expression should be in terms of the velocity of the loop v, the width of the loop w, and the change in time Δt. (*Hint:* Don't forget about the sign of $\Delta \Phi$.)

2. Write an expression for the change in the magnetic flux in terms of the change in area and the magnetic field strength.

3. Combine these two expressions to find an expression for the induced emf in terms of known quantities.

4. How much emf is induced in the loop?

5. How much power is delivered to the lamp?

Reflect

1. How much power would be delivered to the lamp if the dimensions of the loop were 9.00 cm × 10.00 cm?

Section 20-4 Faraday's law explains how alternating currents are generated

Goal: Define the key properties of an ac generator.

Concept Check

1. Explain how an electric generator uses electromagnetic induction to generate an emf.

2. When designing an electric generator, which parameters will improve the power output when increased?

Problem-Solving Review

Thomas needs to design a large AC generator capable of providing an emf with a maximum amplitude of 300 V. The generator will be used to power a 4.00 kW air conditioner. He can use a 0.500 T permanent magnet that leaves room for a circular coil no larger than 1.00 m² and 2000 turns of wire. Let's calculate the final design parameters for the generator.

Set Up

1. Write an expression for the maximum amplitude of the emf provided by the generator in terms of the number of turns, the area of the coil, magnetic field strength, and angular velocity.

2. Rearrange this expression to find an expression for ωA.

3. Write an expression for the resistance of the air conditioner in terms of the average power provided by the generator, number of turns, the area of the coil, magnetic field strength, and angular velocity.

Solve

1. If the area is chosen to be 0.500 m², how large will the angular velocity need to be to provide a maximum emf of 300 V?

2. What is the resistance of the air conditioner?

Reflect

1. How much power could the generator provide if the strength of the magnetic field were to be increased by 10%?

Chapter 21 Alternating-Current Circuits

Section 21-1 Most circuits use alternating current

Goal: Explain the importance of alternating current.

Concept Check

1. Describe qualitatively the difference between mutual inductance and self-inductance.

2. Explain the purpose and function of a transformer.

Section 21-2 We need to analyze AC circuits differently than DC circuits

Goal: Describe what is meant by the root mean square value.

Concept Check

1. Explain in what manner a resistor functions differently in an AC circuit compared to a DC circuit.

2. Explain the difference between the average voltage and the root mean square voltage of an AC circuit.

Problem-Solving Review

In the United States, outlets usually have a root mean square voltage of 120 V. Sometimes for large appliances, outlets are provide with an rms voltage of 240 V. The frequency of the voltage is 60.0 Hz. Let's examine the electrical properties of these higher voltage outlets.

Set Up

1. What is the period of oscillation of the voltage?

2. Graph the voltage as a function of time.

3. Write an expression for the maximum amplitude of the voltage in terms of the rms voltage.

4. Write an expression for the average power in terms of the rms voltage and the resistance of the load plugged in to the outlet.

Solve

1. What is the maximum amplitude of the voltage?

2. A washing machine with a resistance of 30.0 Ω is connected to the outlet. What is the average power used by the washing machine?

Reflect

1. If the frequency were to be doubled, what would the average power used by the washing machine be?

Section 21-3 Transformers allow us to change the voltage of an ac power source

Goal: Calculate the voltage change produced by a transformer.

Concept Check

1. Explain why long distance power lines are designed to use a much higher voltage than the household 120 V.

2. Explain the function of the iron core inside a transformer.

3. Explain how it is possible that the voltage across the primary coil in a transformer maybe different than the voltage across the secondary coil.

Problem-Solving Review

Two transformers are connected together. The primary coil of transformer A is connected to a standard 120 V rms outlet. The secondary coil is connected to the primary coil of transformer B. The secondary coil of B has to provide 30.0 V rms to power a stereo amplifier, which draws 11.7 A. The primary coil of A has 1000 turns, and the secondary of B has 2000 turns. The current in the primary coil of transformer B is measured to be 3.90 A. Let's calculate the number of turns in the other two windings.

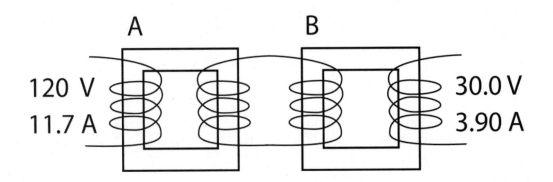

Set Up

1. In the preceding diagram abel the primary coil of the transformer A with the number 1. Label the secondary coil 2. Label the primary coil of the transformer B with the number 3 and the secondary coil 4.

2. What is the relationship between the rms voltage across coil 2 (V_2) and the rms voltage across coil 3 (V_3)?

3. What is the relationship between the power in coil 3 and the power in coil 4?

4. Use this relation to find an expression relating the rms voltage across coil 3, the current through coil 3, the rms voltage across coil 4, and the current through coil 4.

5. Write an expression relating the rms voltage across coil 1, the number of runs in coil 1, the rms voltage across coil 2, and the number of turns in coil 2.

6. Write an expression relating the rms voltage across coil 1, the number of runs in coil 1, the rms voltage across coil 2, and the number of turns in coil 2.

Solve

1. What is the rms voltage across coil 3?

2. What is the rms voltage across coil 2?

3. How many turns are there in coil 3?

4. How many turns are there in coil 4?

Reflect
1. If the input rms voltage across coil 1 were to be doubled, what would the output voltage across coil 4 be?

Section 21-4 An inductor is a circuit element that opposes changes in current

Goal: Describe why an inductor opposes changes in the current passing through it.

Concept Check

1. Explain why the inductance of a coil does not depend on the current flowing through it.

2. Describe the change in voltage across an inductor when the current through it is increasing.

Problem-Solving Review

Karen wants to build her own AM radio. She decides to just duplicate the inductor she found in her stereo, but the coil has too many turns to count. She decides to measure the number of turns by applying a varying current and measuring the voltage output. She measures the diameter of the coil to be 0.491 cm and the length to be 6.13 cm. As Karen applies a current of 2.39 A and then increases the current to 2.95 A in 0.191 ms, she measures a voltage of −0.0458 V.

Set Up

1. What is the cross-sectional area of the coil?

2. Write an expression for the inductance of the coil in terms of the change in current over time and the voltage across the coil.

Chapter 21

3. Write an expression for the number of turns in the coil in terms of the induction, the length of the coil, and the area and the magnetic permeability of free space.

Solve
1. What is the inductance of the coil?

2. How many turns does the coil have?

Reflect
1. How can the inductance of the coil be increased?

Section 21-5 In a circuit with an inductor and capacitor, charge and current oscillate

Goal: Explain the flow of energy in an LC circuit.

Concept Check
1. Describe how current oscillates in an LC circuit.

2. Explain mathematically how the flow of current in an LC circuit is similar to motion of a block on a spring. Describe which quantities in an LC circuit and which quantities in the equation of motion of a block on a spring serve analogous functions.

3. Explain how a microwave oven works.

Problem-Solving Review
A 7.03 µH inductor and a 9.71 µF form an LC circuit with an initial charge on the capacitor of 0.725 µC. Assume the phase angle ϕ is 0.00 rad. Let's explore the electrical properties of this circuit.

Set Up
1. What is angular frequency of this LC circuit?

78 Chapter 21

2. Write an expression for the charge in the capacitor as a function of time and another expression for the current flowing in the circuit as a function of time.

3. Draw a graph of the charge in the capacitor versus time.

4. Draw a graph of the current in the circuit as a function of time.

5. Write an expression for the total electromagnetic energy in the circuit.

Solve
1. With what period does the current oscillate?

2. What is the total energy of the circuit?

3. What is the charge in the capacitor at time $t = 73.4$ μs?

4. What is the current in the inductor at time $t = 153$ μs?

Reflect
1. How can the period of oscillation be increased?

Section 21-6 When an AC voltage source is attached in series to an inductor, resistor, and capacitor, the circuit can display resonance

Goal: Describe what happens in a driven series LRC circuit when the driving frequency changes.

Concept Check

1. Explain how a LRC circuit can display resonance. What happens to the current at very low frequencies? At very high frequencies?

2. Explain the difference between resistance and impedance.

Problem-Solving Review

Sam has a radio with a LRC circuit inside. He needs to adjust a variable capacitor to tune the radio to a frequency of 1050 kHz and he wants to measure the impedance of the circuit. The inductor in the circuit is 7.56 cm long, has a radius of 0.257 cm, and 20,000 turns. The resistor has a value of 1.25 kΩ.

Set Up

1. To what angular frequency should the circuit be tuned so that the radio is tuned to a frequency of 1050 kHz?

2. After the circuit is tuned, what will the natural angular frequency be? What will the driving angular frequency be?

3. What is the inductance of the inductor?

4. Write an expression for the natural angular frequency (which Sam is tuning to the driving frequency of the station) in terms of the inductance and capacitance of the circuit.

5. Write an expression for the impedance of the circuit in terms of the driving angular frequency, the inductance, the capacitance, and the resistance.

Solve
1. To what capacitance should the variable capacitor be tuned to?

2. What is the value of the impedance?

Reflect
1. If Sam tunes the radio perfectly, what will the phase angle ϕ be?

Section 21-7 Diodes are important parts of many common circuits

Goal: Discuss why current can flow in only one direction in a pn junction diode.

Concept Check

1. Explain the difference between p- and n-type semiconductors.

2. Explain the purpose of a diode and how it works.

3. Explain how solar cells work.

Chapter 22 Electromagnetic Waves

Section 22-1 Light is just one example of an electromagnetic wave.

Goal: Define an electro-magnetic wave.

Concept Check

1. Name three types of electromagnetic radiation besides light.

2. What is a photon?

Section 22-2 In an electromagnetic plane wave, electric and magnetic fields both oscillate

Goal: Discuss how speed, frequency, and wavelength are related for electromagnetic waves, and describe the structure of an electromagnetic plane wave.

Concept Check

1. What is the angle between the electric and magnetic field of visible light hitting your eyes?

2. Write an equation to describe the relationship between the magnitude of the electric field and the magnitude of the magnetic field in a sinusoidal electromagnetic plane wave.

Problem-Solving Review

The door to a microwave has a conducting mesh screen with tiny holes that allow visible light to pass, but not microwaves. This is because, in the conductor, the holes are smaller than the wavelength of the microwave, but not the visible light. We will also consider a conducting mesh that lets x-rays pass, but not visible light.

Set Up

1. What is the wavelength of an x-ray with a frequency of 3.00×10^{18} Hz?

2. What is the wavelength of a microwave with a frequency of 5.00×10^{8} Hz?

Solve

1. For a microwave with a mesh screen that has holes with a diameter of 2.00 mm, what is the lowest frequency electromagnetic radiation that can pass?

2. Consider a conducting mesh with holes that have a diameter of 2.00×10^{-9} m. What is the lowest frequency electromagnetic radiation that can pass through this mesh?

3. Can humans detect the frequency in number 2 with their eyes?

Reflect

1. What is the velocity of starlight that arrives at the *Hubble* space telescope?

Section 22-3 Maxwell's equations explain why electromagnetic waves are possible

Goal: Explain what Maxwell's equations are and what they tell us about electromagnetic waves.

Concept Check
1. Write Maxwell's equations, the four basic equations of electromagnetism.

Problem-Solving Review

Consider 2 set ups: The first is a wire running along the x-axis with a current of 2.50 A flowing in the $+x$ direction that produces a circulating magnetic field. The second set up is a parallel plate capacitor where the plates are 0.001 m away from each other. We will find the magnitude of the current charging the parallel plate capacitor that yields an induced magnetic field of similar strength as the magnetic field a distance p from the wire.

Set Up
1. Write the Maxwell ampere law for the wire. Only include non-zero terms.

2. Write the Maxwell ampere law for the charging capacitor. Only include non-zero terms.

3. What is $\Delta \ell$, the circumference of the circle at point p that is a distance of 0.023 m away from the wire?

Solve
1. Use the Maxwell ampere's law for the wire to find the magnetic field a distance p away from the wire.

2. Use your result from the previous question and Maxwell ampere's law for the charging capacitor to find the change in electric flux per second a distance p from the center of the capacitor. Recall that the current charging the capacitor creates a magnetic field of similar strength to the magnetic field around the wire.

3. What is the change in electric field per second caused by the current charging the capacitor?

Reflect
1. Use Gauss's law to equate the change in electric flux previously measured through the loop in the center to the charge that arrives on the capacitor per second.

2. What is the magnitude of the current charging the capacitor?

Section 22-4 Electromagnetic waves carry both electric and magnetic energy, and come in packets called photons

Goal: Calculate the energy density and intensity of an electromagnetic wave and the energy of a photon.

Concept Check
1. The energy of a photon is proportional to what quantity?

2. How is the energy in an electromagnetic wave distributed between electric and magnetic fields?

Problem-Solving Review
The human body emits infrared radiation with a wavelength around 9.51×10^{-6} m. Let's explore how much energy is radiated this way.

Set Up
1. What is the frequency of a photon with a wavelength of 9.51×10^{-6} m?

2. What is the energy per photon for the case found above?

Solve
1. The average body surface area for a person is 1.75 m², and they radiate photons away at around 100 W. What is the intensity, S_{avg}, of the electromagnetic wave radiated away from the person?

2. What is the average energy density, u_{avg}?

3. Use the average energy density to find the electric and magnetic field values, E_{rms} and B_{rms}.

Reflect
1. What are E_o and B_o

Chapter 23 Wave Properties of Light

Section 23-1 The wave nature of light explains much of how light behaves

Goal: Describe some key properties of light.

Concept Check
1. What characteristic of light enables us to describe many of its key features?

Section 23-2 Huygens' principle explains the reflection and refraction of light

Goal: Explain Huygens' principle and what it tells us about the laws of reflection and refraction

Concept Check
1. What is Huygens' principle?

2. What is refraction?

Problem-Solving Review
Consider the following set up:

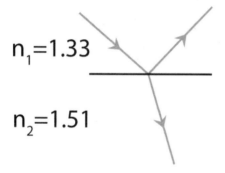

A beam traveling through a medium with $n_1 = 1.33$ is incident on a material with $n_2 = 1.51$. Part of the beam is reflected from the surface, and part is refracted into the second material.

Set Up
1. Make a dashed line on the figure indicating the normal to the surface.

2. Label θ_1, the angle between the incident beam and the normal to the surface. Label θ_2, the angle between the refracted beam and the normal to the surface, and label θ_3, the angle between the reflected beam and the normal to the surface.

Solve
1. If θ_1 is 53.1°, what is θ_2?

2. If θ_1 is 53.1°, what is θ_3?

Reflect
1. What is the speed of the reflected beam?

2. What is the speed of the refracted beam?

Section 23-3 In some cases light undergoes total internal reflection at the boundary between media

Goal: Recognize the special circumstances under which total internal reflection can take place.

Concept Check
1. What is total internal reflection?

2. If light goes from a medium with a lower index of refraction to a medium with a higher index of refraction, does the refracted ray bend toward the normal or away from the normal?

3. If light goes from a medium with a higher index of refraction to a medium with a lower index of refraction, does the refracted ray bend toward the normal or away from the normal?

Problem-Solving Review
Let's consider the total internal reflection of a light source positioned 2.5 m below the surface of a deep pool of water.

Set Up

1. What is the critical angle θ_c for light going from water to air?

Solve

1. Find the maximum radius of a circle on the surface of the water where light from the source can escape from the water.

2. Find the maximum radius for the circle at the surface of the water where light can escape if the light source is now 3 m below the surface.

Reflect

1. If light from the source that is 3 m below the surface is incident on the boundary between water and air at 50° to the normal, what will the angle of reflection be? Will there be refraction?

Section 23-4 The dispersion of light explains the colors from a prism or a rainbow

Goal: Explain how a prism is able to break white light into its component colors.

Concept Check
1. What is dispersion?

2. For most materials, is the index of refraction higher for red light or blue light?

Problem-Solving Review
Consider a ray of light incident on a material with an unknown index of refraction that is $w = 57.8$ mm thick. When white light is incident on the material at an angle of 26° with respect to the normal, red and blue light is refracted at the boundary. The red light illuminates the bottom of the material at a distance $d_{red} = 17.4$ mm from the intersection of the bottom of the material and a line drawn from the normal to the material's surface. The blue light illuminates the bottom of the material at a distance of $d_{blue} = 17.7$ mm from the normal.

Set Up
1. For the red light, write the distance the beam travels through the material, h_r, in terms of w, the thickness of the material, and d, the distance of the exiting beam from the normal. Solve for h_r.

2. What is θ_r?

3. For the red light, write the distance the beam travels through the material, h_b, in terms of w, the thickness of the material, and d, the distance of the exiting beam from the normal. Solve for h_b.

4. What is θ_b?

Solve

1. What is n_{red}? (Assume the environment surrounding the material has an index of refraction of 1.00.)

2. What is n_{blue}?

Reflect

1. What is the speed of red light in the material?

2. What is the speed of blue light in the material?

Section 23-5 In a polarized light wave, the electric field vector points in a specific direction

Goal: Calculate how the intensity of light is affected by passing through a polarizing filter.

Concept Check

1. What are three ways light can be polarized?

2. What is unpolarized light?

Problem-Solving Review

Your boss wants you to design a series of two polarized light filters such that any unpolarized light that passes through the series will leave at 1/3 its original intensity, I_o = 351 W/m². The first filter needs to be at an angle of 37.0° from the vertical.

Set Up

1. What is the polarization of the light after it goes through the first filter?

2. What is the intensity of the light after it goes through the first filter?

Solve

1. What is the difference in intensity after the light passes between the first filter and $I_o/3$, the desired final intensity?

2. Find the angle for the second filter with respect to the first filter that will yield the correct intensity.

Reflect
1. What is the polarization of the light leaving the filter?

Section 23-6 Light waves reflected from the layers of a thin film can interfere with each other, producing dazzling effects

Goal: Use the idea of path length difference to calculate what happens in thin-film interference.

Concept Check
1. What is thin film interference?

Problem-Solving Review
While working in a kitchen someone spills some corn oil with an index of refraction of 1.47. Light with a frequency of 577×10^{12} Hz is shone on the oil slick and destructive interference is observed.

Set Up
1. What is the wavelength of the incident beam?

Solve
1. What is the minimum thickness of the oil for destructive interference?

2. What are two other thickness values the oil could have that would cause the same effect?

3. What are three thickness values for the oil that would cause constructive interference?

Reflect

1. What is the minimum path length difference, Δ_{pl}, for the destructive interference case?

2. What is the minimum path length difference, Δ_{pl}, for the constructive interference case?

Section 23-7 Diffraction is the spreading of light when it passes through a narrow opening

Goal: Explain why light spreads out when it passes through a narrow opening.

Concept Check
1. Why does light spread out when it passes through a narrow opening?

Problem-Solving Review
A diffraction grating is a surface with numerous small slits cut into it. In an experiment light from a mercury lamp is incident on a diffraction grating with 15,000 slits per inch at the normal to the orientation of the grating. The light passes through the grating and then lands on a flat surface $d = 0.15$ m away.

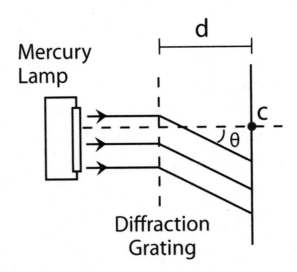

Using this information, it is possible to predict how far away light of each wavelength emitted from the mercury lamp will land on the surface from point c.

Color of light after passing through the grating	Wavelength (nm)
violet	405
blue	436
green	546
yellow	578

Set Up
1. What is d, the slit width?

Solve
1. What is the angle between the normal to the grating and the direction of the violet and blue light after it passes through the grating?

2. What is the angle between the normal to the grating and the direction of the green and yellow light after it passes through the grating?

3. How far away from point c is the first instance of violet and blue light?

4. How far away from point c is the first instance of green and yellow light?

Reflect

1. How far away from point c is the second violet fringe?

2. How far away from point c is the third blue fringe?

Section 23-8 The diffraction of light through a circular aperture is important in optics

Goal: Calculate how the angular resolution of an optical device is limited by diffraction.

Concept Check
1. What is Rayleigh's criterion?

2. Describe what angular resolution is.

Problem-Solving Review
You wish to determine a microscope's average resolution for visible light. This can be done by finding Rayleigh's criterion for blue light and red light and then averaging them together. The area of the aperture is $5.12 \times 10^{-4}\,\text{m}^2$. The blue light has a wavelength of 405 nm, and the red light has a wavelength of 702 nm.

Set Up
1. What is the angular resolution for blue light?

2. What is the angular resolution for red light?

Solve

1. What is the smallest distance away from the microscope two point sized objects that are 1.01 mm away from each other can be resolved with blue light?

2. What is the smallest distance away from the microscope two point sized objects that are 1.01 mm away from each other can be resolved with red light?

Reflect

1. What is the difference in angular resolution between the blue light and red light?

2. What is the average angular resolution for the microscope?

Chapter 24 Geometrical Optics

Section 24-1 Mirrors or lenses can be used to form images

Goal: Explain the importance of optical devices

Concept Check

1. Name three optical devices.

2. What physical phenomenon do lenses rely on to work?

3. What physical phenomenon do mirrors rely on to work?

Chapter 24

Section 24-2 A plane mirror produces an image that is reversed back to front

Goal: Describe how a plane mirror forms an image.

Concept Check
1. What makes an image real or virtual?

2. Compared to the direction of light incident on a plane mirror, in what direction will reflected rays travel?

Problem-Solving Review
An object with height $h_o = 0.121$ m is held up 1.00 m in front of a plane mirror.

Set Up
1. Sketch the object in front of the mirror.

2. What is the object distance, d_o?

Solve

1. On your sketch in Question 1 of the Set Up, sketch a light ray from the top of the object to the mirror. Also sketch the reflected ray.

2. What is the image distance, d_i?

3. What is the image height, h_i?

Reflect

1. What is the magnification, m?

Section 24-3 A concave mirror can produce an image of a different size than the object

Goal: Use ray diagrams to explain how the image formed by a concave mirror depends on the position of the object.

Concept Check

1. If an object is beyond the focal length of a concave mirror, will its image be upright or inverted? Magnified or unmagnified?

2. If an object is less than the focal length of a concave mirror, will its image be upright or inverted? Magnified or unmagnified?

3. What is the image size of an object at the focal point of a concave mirror?

Problem-Solving Review

Case A: object ↑, C, f,) — — — Principal Axis

Case B: — — C — — f ↑ image) — — — Principal Axis

Set Up

1. For case A, draw a ray that goes directly to the mirror and through the focal point. Also draw the ray that goes to the principle axis. Where these rays intersect, draw an arrow—that's where the image is. If they don't intersect, draw dashed lines behind the mirror to indicate where the imaginary image will be.

2. For case B, draw a ray that goes directly to the mirror and through the focal point. Also draw the ray that goes to the principle axis. Where these rays intersect, draw an arrow—that's where the image is. If they don't intersect, draw dashed lines behind the mirror to indicate where the imaginary image will be.

Solve

1. Circle one from each column that applies to the image formed in case A.

real	upright	image appears larger
imaginary	inverted	image appears smaller

2. Circle one from each column that applies to the image formed in case B.

real	upright	image appears larger
imaginary	inverted	image appears smaller

Reflect

1. What is the relationship between the focal point and the radius of curvature?

Section 24-4 Simple equations give the position and magnification of the image made by a concave mirror

Goal: Calculate the position and height of an image made by a concave mirror.

Concept Check

1. What does a positive value of d_i indicate about the image reflected from a concave mirror?

2. What does a negative value of d_i indicate about the image reflected from a concave mirror?

Problem-Solving Review

An object with a height of 0.197 m is 0.774 m away from a concave mirror that has a radius of curvature of 1.23 m.

Set Up

1. What is the focal length of this mirror?

Solve

1. What is the image distance, d_i?

2. What is the magnification of the object?

Reflect

1. What is the height of the object?

Section 24-5 A convex mirror produces an image that is smaller than the object.

Goal: Explain the differences between the images made by a convex mirror and a concave mirror.

Concept Check
1. Do reflected light rays converge or diverge for a concave mirror?

2. Do reflected light rays converge or diverge for a convex mirror?

Problem-Solving Review

a ---- Principal Axis

b ---- Principal Axis

Set Up

1. For the convex mirror in case A, draw a ray that goes directly to the mirror and reflects. Also draw the ray that goes to the principle axis and its reflected ray. Draw dashed lines behind the mirror to indicate where the imaginary image will be.

2. For the convex mirror in case B, draw a ray that goes directly to the mirror and reflects. Also draw the ray that goes to the principle axis and its reflected ray. Draw dashed lines behind the mirror to indicate where the imaginary image will be.

Solve

1. Circle one from each column that applies to the image formed in case A.

real	upright	image appears larger
imaginary	inverted	image appears smaller

2. Circle one from each column that applies to the image formed in case B.

real	upright	image appears larger
imaginary	inverted	image appears smaller

Reflect

1. Will the image appear larger in case A or case B?

2. Will the magnitude of the focal point for both case A and case B be positive or negative?

Section 24-6 The same equations used for concave mirrors also work for convex mirrors

Goal: Calculate the position and height of an image made by a concave mirror.

Concept Check
1. What is the difference between a mirror with a positive radius of curvature and a mirror with a negative radius of curvature?

Problem-Solving Review
An object with a height of 0.372 m is 1.54 m away from a convex mirror. The image created appears to be 0.781 m behind the mirror.

Set Up
1. What is the image distance, d_i?

Solve
1. What is the focal length of the mirror?

2. Use the image distance, d_i, and object distance, d_o, to find the magnification of the mirror.

3. What is the image height, h_i?

Reflect

1. What is the radius of curvature of the mirror? Is it positive or negative?

Section 24-7 Convex lenses form images like concave mirrors and vice versa

Goal: Describe how the curved surfaces of a lens make light rays converge or diverge.

Concept Check
1. What physical phenomenon causes light passing through a lens to converge or diverge?

Problem-Solving Review

A – – ↑– –•– –)(– – –•–
         f   f

B – – – –•↑– –)(– – –•–
              f   f

C – –↑– –•– –)|(– – –•–
         f   f

Set Up

1. For the convex lens with $d_o > f$, draw three rays from the object through the lens: one from the object parallel to the principle axis that goes through the focal point after being refracted, one that goes from the object to the principle axis, and one that goes through the focal point and parallel to the principle axis after being refracted by the lens.

2. For the convex lens with $f > d_o$, draw two rays from the object through the lens: one from the object parallel to the principle axis that goes through the focal point after being refracted, and one from the object to the principle axis. Be sure to draw dashed imaginary lines back from the refracted rays that show the image position.

2. For the concave lens, draw two rays from the object through the lens: one from the object traveling parallel to the principle axis and refracted by the lens, and one from the object through the principle axis. Be sure to draw dashed imaginary lines back from the refracted rays that show the image position.

Solve

1. Circle one from each column that applies to the image formed in case A.

real	upright	image appears larger
imaginary	inverted	image appears smaller

2. Circle one from each column that applies to the image formed in case B.

real	upright	image appears larger
imaginary	inverted	image appears smaller

3. Circle one from each column that applies to the image formed in case C.

real	upright	image appears larger
imaginary	inverted	image appears smaller

Reflect

1. What happens to the image distance for $d_o = f$?

Section 24-8 The focal length of a lens is determined by its index of refraction and the curvature of its surfaces

Goal: Calculate the focal length of a lens based on its composition and shape.

Concept Check
1. What two factors determine the location of the focal point of a thin lens?

Problem-Solving Review
You are designing a thin lens made of acrylic that has an index of refraction of 1.49. An object placed 0.851 m in front of the lens is required to make an image at $d_i = -0.234$ m. The piece of acrylic that you will make the thin lens from has one side precut to a radius of curvature $R_2 = 0.228$ m.

Set Up
1. What is the required focal length of the lens?

Solve
1. Find R_1 for the lens that will yield the correct specifications.

2. What is the magnification?

122 Chapter 24

3. If the object has height of 0.100 m, what is the image height?

Reflect

1. Is the image real or imaginary?

Chapter 25 Relativity

Section 25-1 The concepts of relativity may seem exotic, but they're part of everyday life

Goal: Identify some circumstances under which the physics you know breaks down.

Concept Check

1. What is an example of an everyday device that requires an understanding of relativity to properly function?

2. What quantity related to an object's motion can be changed so that the physics learned so far needs to be modified?

Section 25-2 Newton's mechanics includes some ideas of relativity

Goal: Describe how different observers view the same motion in Newtonian physics.

Concept Check
1. What type of transformation allows the transfer of velocities and coordinates between inertial reference frames?

Problem-Solving Review
Sitting next to a sidewalk watching people go by, you measure person A's velocity at 1.56 m/s in the +x direction and person B's velocity at 2.98 m/s in the +x direction.

Set Up
1. What would person B measure person A's velocity to be?

2. What would person A measure person B's velocity to be?

Solve
1. What would person A say your velocity was?

2. What would person B say your velocity was?

Reflect

1. Is there any "absolute" reference frame between you, person A, and person B?

Section 25-3 Einstein's relativity predicts that the time between events depends on the observer

Goal: Explain how the Michelson-Morley experiment helped rule out the ether model of the propagation of light.

Concept Check

1. Can sound waves propagate in a vacuum?

2. The last step of the Michelson-Morley experiment has the two recombined beams forming an interference pattern on a screen. With no ether, the beams arrive at the same time and there is constructive interference. What were Michelson and Morley expecting to happen to the two recombined beams as the apparatus moved through the ether?

Section 25-4 Einstein's relativity predicts that the time between events depends on the observer

Goal: Describe how the time interval between two events can have different values in different frames of reference.

Concept Check

1. If you are watching someone walk in the +x direction at a speed of 3.00 m/s and they shine a flashlight in the +x direction, what would the Newtonian picture predict the velocity of the light ray to be?

2. If you are watching someone walk in the +x direction at a speed of 3.0 m/s and they shine a flashlight in the +x direction, what would the relativistic picture predict the velocity of the light ray to be?

Problem-Solving Review

Erin boards a 6-hour flight to France and sets her watch to match her twin sister Lauren's watch. A typical commercial airplane travels at around 500 mph. The sisters will compare the times on their watches when Erin arrives in France.

Set Up

1. What is the velocity of the plane in m/s?

Solve

1. What is Δt, the time of the plane's trip measured by Lauren in the stationary reference frame?

2. What is Δt_{proper}, the time of the plane's trip measured by Erin in the plane's reference frame? For this problem, ignore the typical significant-figure rules.

3. What is the difference in times that they measured?

Reflect

1. What velocity is necessary for Erin to measure $\Delta t_{proper} = 0.99\, \Delta t$?

Section 25-5 Einstein's relativity also predicts that the length of an object depends on the observer

Goal: Calculate how the dimensions of an object change when it is in motion.

Concept Check
1. What is length contraction?

Problem-Solving Review
If you ran very quickly toward a barn with a length of 2.00 m with a pole that measures 10.0 m long in a reference frame at rest with respect to the pole, it is possible that in the reference frame of the barn it would appear that the pole was entirely inside the barn for a brief period due to length contraction.

Set Up
1. Take S' to be the reference frame of the runner with the pole. What is the length of the pole in S'?

Solve
1. How fast would an observer in reference frame S say you would have to run for the 10.0-m-long pole in reference frame S' to appear 2.00 m long in reference frame S?

2. In reference frame S', what is the velocity of the barn?

3. What is the length of the barn in reference frame S'?

Reflect

1. In S, the pole would appear entirely inside the barn (albeit briefly). Describe qualitatively what an observer in S' would report about the pole fitting inside the barn.

Section 25-6 The relative velocity of two objects is constrained by the speed of light, the ultimate speed limit

Goal: Explain why the speed of light in a vacuum is an ultimate speed limit.

Concept Check
1. What determines whether you can use a Galilean transformation or a Lorentzian transformation?

Problem-Solving Review
A space ship can travel at 0.800 c with respect to reference frame S and launch an escape pod with a velocity of 0.201 c, as witnessed by an observer on the ship.

Set Up
1. Sketch the situation above, and label S the frame watching the rocket go by, S' the frame of the rocket, V the velocity of S' in S, and v_x' the velocity of the escape pod in the S' reference frame.

Solve
1. What is the velocity of the escape pod with respect to an observer in S, v_x?

2. If $v_x' = 0.800$ c, what would v_x be?

Reflect

1. What would v_x be for the case where $v_x' = 0.201$ c using a Galilean transformation, and why would using that type of transformation be correct?

Section 25-7 The equations for kinetic energy and momentum must be modified at very high speeds

Goal: Calculate the rest energy of an object with mass.

Concept Check
1. What would happen to an accelerating particles kinetic energy as you accelerated it closer and closer to c?

Problem-Solving Review
At a particle accelerator, gold ions (the nucleus of a gold atom) with a resting mass of 3.27×10^{-25} kg are accelerated from rest until they have a kinetic energy of 6.47×10^{-8} J.

Set Up
1. What is the rest energy of the gold ion?

Solve
1. What is the relativistic gamma for the accelerated particle?

2. What is the velocity of each gold ion as measured from the reference frame of the collider?

3. What is the momentum of each gold ion as measured from the reference frame of the collider?

Reflect

1. Use the velocity calculated in Question 2 to find the classical momentum of the particle.

Section 25-8 Einstein's general theory of relativity describes the fundamental nature of gravity

Goal: Explain what the principle of equivalence tells us about the nature of gravity.

Concept Check

1. What is the principle of equivalence?

2. What does the principle of equivalence tell us about the nature of gravity?

3. What is an example of light bending in a gravitational field observed in nature?

Chapter 26 Quantum Physics and Atomic Structure

Section 26-1 Experiments that probe the nature of light and matter reveal the limits of classical physics

Goal: Recognize the limitations of classical physics for explaining the properties of light and matter.

Concept Check

1. Explain what it means for an atom's energy to be quantized.

Section 26-2 The photoelectric effect and blackbody radiation show that light is absorbed and emitted in the form of photons

Goal: Describe how the photoelectric effect and blackbody radiation provide evidence for the photon picture of light.

Concept Check

1. Describe the photoelectric effect and explain why it suggests electromagnetic radiation is not only a wave.

2. Explain how photoemission electron microscopy works.

3. Explain why you usually do not see blackbody radiation from objects at room temperature.

4. Explain how experiments with blackbody radiation supported the idea of photons.

Problem-Solving Review

Isaac is working in the lab with a piece of aluminum that has a work function of 6.59×10^{-19} J. Let's calculate which frequency Isaac should tune his laser to eject electrons from the metal via the photoelectric effect with a maximum kinetic energy of 1.60×10^{-18} J.

Set Up

1. Write an expression relating the maximum kinetic energy of electrons emitted from the surface, the frequency of the incoming photons and the work function of the material.

2. Write an expression relating the energy of a photon to its frequency.

Solve
1. Rearrange the expression from Question 1 of the Set Up, solving for the frequency of the incoming photons.

2. What frequency should Isaac tune his laser to?

3. What will the minimum wavelength of the ejected photons be?

Reflect
1. How much energy does one of the photons in the laser have?

Problem-Solving Review
A piece of metal emits 13.2 watts of radiation when it is heated to 100° Fahrenheit. Let's calculate how high the temperature needs to be for the metal to radiate 39.6 watts of radiation.

Set Up
1. What is 100° Fahrenheit in Kelvin?

2. Write an expression relating the power radiated by the metal to its temperature. Write this expression twice, once for the power radiated when the metal is it is initial temperature and once for when the metal has been heated enough to radiate 39.6 watts. For the first case, label the power P_1 and

the temperature T_1. For the second case, label the power P_2 and the temperature T_2.

Solve
1. Write an expression for the ratio of P_2 to P_1.

2. Rearrange this expression to find the ratio of T_2 to T_1.

3. How hot does the metal need to be to emit 39.6 watts of radiation?

Reflect
1. Is the average frequency of emitted radiation higher when the metal is at temperature T_1, T_2, or the same in both cases?

Section 26-3 As a result of its photon character, light changes wavelength when it is scattered

Goal: Explain why the wavelength of a photon increases if it scatters from an electron.

Concept Check

1. Explain how a photon's momentum relates to its energy and frequency.

2. Describe what happens during Compton scattering. What does this tell us about the nature of photons?

Problem-Solving Review

A photon with an energy of 9.13×10^{-14} J Compton scatters off an electron at rest. The energy of the scattered photon is measured to be 3.17×10^{-14} J. Let's calculate the angle at which the photon scattered.

Set Up

1. Write an expression relating the wavelength of a photon to its energy.

2. Write an expression for the change in wavelength of a photon undergoing Compton scattering.

3. Solve this expression for the angle with which the photon scatters.

Solve
1. What is the initial wavelength of the photon? What is the final wavelength?

2. At what angle did the photon scatter?

Reflect
1. If the photon were to scatter at a smaller angle, would it lose less energy, more energy, or the same amount of energy?

Chapter 26

Section 26-4 Matter, like light, has aspects of both waves and particles

Goal: Calculate the wavelength of a particle such as an electron.

Concept Check

1. Explain how it can be shown experimentally that electrons have a wave nature in addition to their particle nature.

2. Explain why we do not normally observe wave-like behavior from everyday objects.

Problem-Solving Review

Will is designing an electron microscope that needs to be able to resolve structures of at least 0.0500 nm or smaller. Let's calculate what the voltage applied to each electron should be.

Set Up

1. Write an expression relating the change in electric potential energy of the electron to the voltage applied.

2. Write an expression relating the momentum of the electron to its wavelength.

Solve

1. Calculate the minimum amount of momentum the electron needs to have to probe length scales of 0.0500 nm.

2. If the mass of the electron is 9.11 × 10⁻²⁸ g, what is its minimum velocity?

3. What is the minimum kinetic energy of the electron?

4. How much voltage should be applied to the electron to reach this energy?

Reflect

1. In order to probe smaller objects, should the energy of the electrons be increased or decreased?

Section 26-5 The spectra of light emitted and absorbed by atoms show that atomic energies are quantized

Goal: Describe why atoms absorb light at only certain wavelengths and why they emit and absorb light at the same wavelengths.

Concept Check

1. Explain the Rutherford experiment and what it revealed about atomic structure.

2. Explain what absorption and emission spectra are and what they reveal about atomic structure.

Problem-Solving Review

A hydrogen atom is observed emitting light at a wavelength of 486.01 nm. The electron that emitted the photon started out in the $n = 4$ energy level. Let's calculate final energy levels occupied by the electron in the atom.

Set Up

1. Find an expression relating the wavelength of emitted light to the energy levels the electron fell to and from.

Solve

1. Solve this expression for the final energy level of the electron.

2. What is the final energy level of the electron?

Reflect
1. What is the energy of the emitted photon?

Section 26-6 Models by Bohr and Schrödinger give insight into the intriguing structure of the atom

Goal: Explain how the Bohr model of the atom explains the spectrum of hydrogen.

Concept Check

1. Explain why electron orbits around hydrogen are quantized in multiples of $\hbar$.

2. Explain what it means for an orbiting electron to have negative energy.

3. Explain the Franck-Hertz experiment and its consequences for the energy levels of electrons in atoms with more than one electron.

4. Explain the Pauli exclusion principle and its consequences for multi-electron atoms.

Problem-Solving Review

Let's explore the structure of a helium ion that has only one electron, but a nucleus of two protons.

Set Up

1. What is the charge of the nucleus? What is Z for the helium ion?

2. Find an expression for the energy of the electron in different orbits in terms of Z and the ground-state energy level of hydrogen E_0.

3. Find an expression for the radius of the electron's orbit in terms of the Bohr radius and Z.

4. Find an expression for the wavelength of the electron in terms of the radius of its orbit.

Solve

1. What is the energy of the electron in eV when it is in the 3rd orbital?

2. What is the radius of the electron's orbit when it is in the 2nd energy level?

3. What is the wavelength of the electron when it is in the 2nd energy level?

Reflect

1. Explain why the Bohr model fails for atoms with more than one electron.

Chapter 27 Nuclear Physics

Section 27-1 The quantum concepts that help explain atoms are essential for understanding the nucleus

Goal: Explain why quantum ideas play an important role in nuclear physics.

Concept Check

1. Explain which fundamental forces are present in the nucleus.

2. Explain the difference between nuclear fission and nuclear fusion.

Section 27-2 The strong nuclear force holds nuclei together

Goal: Describe how we know that the force that holds the nucleus together is both strong and of short range.

Concept Check

1. Describe the properties of the strong nuclear force.

2. Describe the process that leads to nuclear decay.

3. Explain why larger nuclei tend to have more neutrons than protons.

4. Describe how an MRI machine works.

Problem-Solving Review

Let's explore the properties of the following nuclides: tritium ^{3}H, gold ^{197}Au, radium ^{226}Ra. Assume $r_0 = 1.2$ fm and the average mass of a nucleon is 1.67×10^{-27} kg.

Set Up

1. Write an expression relating the radius of a nucleus to its atomic mass number.

2. Write an expression for the volume of a sphere in terms of its radius.

3. Write an expression for the density of an object in terms of its mass and volume.

Solve
1. Estimate the radius of each nuclide.

2. Estimate the volume of each nuclide.

Reflect
1. Estimate the density in kg/m³ of each nuclide.

Section 27-3 Some nuclei are more tightly bound and more stable than others

Goal: Explain how and why the binding energy per nucleon in a nucleus depends on the size of the nucleus.

Concept Check

1. Explain why the mass of a nucleus is less than the sum of the masses of its constituent protons and neutrons.

2. Explain what nuclear binding energy is.

3. Explain why, as function of atomic mass number A, the nuclear binding energy per nucleon increases up to 8.8 MeV and then decreases again.

4. Explain the relationship between a nucleus' binding energy and its stability.

Problem-Solving Review

Uranium ^{235}U is a common fuel used in nuclear reactors. Let's calculate the radius and binding energy of ^{235}U. The mass of ^{235}U is 235.0439 u, the mass of the proton is approximately 938.27 MeV/c^2, and the mass of the neutron is 939.57 MeV/c^2. Uranium has an atomic number of 92.

Set Up

1. What is the mass of ^{235}U in MeV/c^2?

152   Chapter 27

2. Write an expression relating the radius of a nucleus to its atomic mass number.

3. Write an expression for the binding energy of a nucleus.

Solve
1. Estimate the radius of ^{235}U.

2. What is the binding energy of ^{235}U.

Reflect
1. If a ^{235}U nucleus undergoes nuclear fission, will the decay products have a smaller or larger binding energy than the ^{235}U nucleus?

Nuclear Physics

Section 27-4 The largest nuclei can release energy by undergoing fission and splitting apart

Goal: Calculate the energy released in nuclear fission.

Concept Check
1. Explain how energy is released in nuclear fission.

2. Explain how a chain reaction is created.

Problem-Solving Review
Let's complete the following nuclear fission reaction and calculate the energy released by it:

$$^{235}U + n = {}^{141}Ba + ? + 3n$$

Set Up
1. How many protons are on the left side of the reaction equation? How many are on the right side not including the missing reactant?

2. How many protons does the missing reactant have? What is the symbol of this element?

3. How many neutrons does the missing reactant have? Write the symbol of this isotope, including the atomic mass number.

4. Write an expression for the binding energy of a nucleus.

Solve
1. Calculate the binding energies of all of the reactants in the equations. The mass of ^{235}U is 235.0439 u, ^{141}Ba is 140.9165 u, and the mass of the missing reactant is 91.9262 u.

2. How much energy is released in this reaction?

Reflect
1. How does this amount of energy compare to the binding energy of the krypton nucleus?

Section 27-5 The smallest nuclei can release energy if they are forced to fuse together

Goal: Describe why very high temperatures are needed for nuclear fusion reactions.

Concept Check

1. Explain why small nuclei are needed for nuclear fusion.

2. Describe the proton-proton cycle in the Sun. explain each step in detail.

3. Explain why it takes so long for a star to cool down.

Problem-Solving Review

In very hot stars it is possible that 3 ^{3}He nuclei fuse to form a ^{12}C nucleus. Let's calculate the amount of energy released in this reaction. A ^{3}He nucleus has an atomic mass of 3.0160 u, and ^{12}C has a mass of 12.0000 u.

Set Up

1. What is the mass of a ^{3}He nucleus in MeV/c^2? A ^{12}C nucleus?

2. How many neutrons does a ^{3}He nucleus have? Protons?

3. How many neutrons does a ^{12}C nucleus have? Protons?

4. Convert 1 MeV to Joules.

Solve
1. Calculate the binding energy of a ^{3}He nucleus.

2. Calculate the binding energy of a ^{12}C nucleus.

2. How much energy is released in this reaction?

Reflect
1. How many reactions per second would need to take place to provide 1 kW of power?

Section 27-6 Unstable nuclei may emit alpha, beta, or gamma radiation

Goal: Calculate what happens in the decay of a radioactive substance.

Concept Check

1. Describe what it means for something to be radioactive.

2. Describe what the decay rate and the decay probability λ are.

3. Explain what the nuclear half-life is.

4. Explain the processes that produce beta and gamma radiation.

Problem-Solving Review

^{228}Th decays into ^{224}Ra + α with a half-life of 1.91 years. Let's calculate how long it takes a 16.0-gram sample of ^{228}Th to decay into 0.0625 g of ^{224}Ra, the decay probability, and how much energy is released in this decay. The mass of ^{228}Th is 228.02874 u, the mass of ^{224}Ra is 224.02021 u, and the mass of ^{4}He is 4.0026033 u.

Set Up

1. Find an expression relating the half-life of a radioactive substance to its decay probability.

2. In what number of half-lives will a 16.0-gram sample of ^{228}Th decay into 0.0625 g of ^{224}Ra?

3. How many neutrons does ^{228}Th have? Protons?

4. How many neutrons does ^{224}Ra have? Protons?

5. How many neutrons does ^{4}He have? Protons?

6. Calculate the masses of ^{228}Th, ^{224}Ra, and ^{4}He in MeV/c^2.

Solve

1. Calculate how long it takes a 16.0-gram sample of ^{228}Th to decay into 0.0625 g of ^{224}Ra.

2. Calculate the decay probability λ.

3. Calculate the binding energies of ^{228}Th, ^{224}Ra, and ^{4}He.

4. How much energy is released in this decay?

Reflect

1. Instead of 16.0 g, imagine having 8.00×10^{23} ^{228}Th atoms. Write an expression for the rate of decay as a function of time.

Chapter 28 Particle Physics and Beyond

Section 28-1 Studying the ultimate constituents of matter helps reveal the nature of the universe

Goal: Explain why physicists examine both the smallest and largest objects in the universe.

Concept Check
1. What kinds of matter are studied in particle physics?

2. Name some fundamental particles.

Particle Physics and Beyond

Section 28-2 Most forms of matter can be explained by just a handful of fundamental particles

Goal: Describe the difference between hadrons and leptons, and explain what hadrons are made of.

Concept Check

1. Explain the difference between quarks and leptons.

2. Explain how we know that hadrons are made up of quarks.

3. Explain how a neutrally charged neutron can possess a magnetic field.

4. Describe the effect of quark confinement.

5. Explain what antimatter is.

6. Explain what between hadrons, baryons, and mesons are.

Problem-Solving Review

Let's examine a few possible processes in particle physics and see if they are possible.

Set Up

1. Fill in the baryon and lepton numbers for the following particles:

Symbol	Baryon number	Electron-lepton number	Muon-lepton number	Tau-lepton number
p				
p̄				
n				
n̄				
e^+				
e^-				
μ^-				
μ^+				
π				
π				
τ^-				
τ^+				
ν_e				
$\bar{\nu}_e$				
ν_μ				
$\bar{\nu}_\mu$				
ν_τ				
$\bar{\nu}_\tau$				
D_s^+				
Γ				

Solve

1. For each of the following interactions, state whether the interaction is possible. If it is not possible, explain which conversation law is violated.

   i. $D_s^+ \to \mu + \nu_\mu$

   ii. $D_s^+ \to \tau + \nu_\tau + n$

   iii. $\tau \to e^- + \nu_\tau$

   iv. $\tau \to \pi^- + 3\pi^0 + \nu_\tau$

   v. $\tau \to \pi^- + 3\pi^0 + \nu_\tau + n + n$

   vi. $e^- + e^+ \to 2\gamma$

Reflect

1. What other conservation laws besides baryon and lepton numbers will contribute to whether or not a certain interaction is possible?

Section 28-3 Four fundamental forces describe all interactions between material objects

Goal: Explain how fundamental particles interact with each other, and the differences among the four fundamental forces.

Concept Check

1. Explain Heisenberg's uncertainty principle and its consequences for the conservation of energy.

2. Explain how forces arise and what virtual particles are.

3. Describe what gluons are.

4. Explain why it is impossible to isolate a quark.

5. Explain why the strong nuclear force has such a short range.

6. Describe what the Standard Model is.

Problem-Solving Review

Imagine a hypothetical fifth fundamental force that has a range of 2.25 fm. Let's estimate the mass of this force's mediating particle.

Set Up
1. Write an expression for Heisenberg's uncertainty principle in terms of the uncertainty in energy and time.

2. Write an expression for the maximum distance the mediating particle can travel in terms of time and the speed of light.

3. Write an expression for the rest energy of the meditating particle in terms of its rest mass.

Solve
1. Solve Heisenberg's uncertainty principle for the uncertainty in time Δt.

2. Substitute this expression for the time t in the equation you found in Question 2 of the Set Up. Also, for the energy E substitute the expression you found in Question 3 of the Set Up.

3. Solve this equation for the mass of the mediating particle.

4. Find the mass of the mediating particle in MeV/c^2.

Reflect

1. For a longer range, would a mediating particle with a smaller or larger mass be needed?

Section 28-4 We live in an expanding universe, and the nature of most of its contents is a mystery

Goal: Calculate the distance to a remote galaxy from its recessional velocity.

Concept Check

1. Explain how we know that the universe is expanding.

2. Explain how we know there was a big bang.

3. Describe what the cosmic microwave background is.

4. Explain why we think there is dark matter in the universe.

Problem-Solving Review

The Andromeda Galaxy is about two and a half million light-years away. Let's calculate how fast the Andromeda Galaxy is receding.

Set Up

1. Solve Hubble's law for the velocity of a receding galaxy.

Solve
1. Approximately how fast is the Andromeda Galaxy receding?

Reflect
1. How long ago was the light that we are observing now, emitted from the Andromeda Galaxy?